GARDEN
BIRDS
of
DELHI, AGRA & JAIPUR

*'A bird does not sing because
it has an answer.
It sings because it has a song.'*

– Chinese proverb

GARDEN BIRDS
of
DELHI, AGRA & JAIPUR

Text
Samar Singh

Photographs
Nikhil Devasar
Amano Samarpan

Photo credits
Nikhil Devasar: half title page, full title page, back cover, and pp. viii, 2, 4, 16-28, 34, 36, 46-56, 60, 64, 68, 72, 80, 82, 84, 87, 89, 91, 93, 95, 97, 99, 101.
Amano Samarpan: cover page, and pp. 8-14, 30, 32, 38-42, 58, 62, 66, 70, 74, 76, 86.

© India International Centre, 2007

First published 2008
Co-published with India International Centre
(expanded version of *Birds of IIC*)

Project coordinator and editor
Bela Butalia

All rights reserved. No part of this publication may be reproduced or transmitted, in any form or by any means, without prior permission of the author and the publisher.

ISBN 13: 978-81-8328-076-1
ISBN 10: 81-8328-076-5

Published by
Wisdom Tree
4779/23 Ansari Road
Darya Ganj
New Delhi-110002
Ph.: 23247966/67/68

Published by Shobit Arya for Wisdom Tree; *typeset at* Marks & Strokes, Delhi-110002 and *printed at* AAR VEE Printers Pvt. Ltd. New Delhi-110064

Contents

Foreword	*vii*
Dr Karan Singh	
Nehru's Letter	*viii*
Introduction	*ix*
BIRD DESCRIPTIONS AND PLATES	
Little Green Bee-eater–*Merops orientalis*	1
Oriental White-eye–*Zosterops palpebrosus*	3
Purple Sunbird–*Nectarinia asiatica*	5
Blue Rock *Thrush–Monticola solitarius*	7
Oriental Magpie Robin–*Copsychus saularis*	9
Asian Paradise Flycatcher–*Terpsiphone paradisi*	11
House Swift–*Apus affinis*	13
Red-vented Bulbul–*Pycnonotus cafer*	15
White-throated Kingfisher–*Halcyon smyrnensis*	17
Common Kingfisher–*Alcedo atthis*	19
Spotted Owlet–*Athene brama*	21
Common Tailorbird–*Orthotomus sutorius*	23
Yellow-crowned Woodpecker–*Dendrocopos mahrattensis*	25
House Sparrow–*Passer domesticus*	27
Indian Robin–*Saxicoloides fulicata*	29
Golden Oriole–*Oriolus oriolus kundoo*	31
Baya Weaver–*Ploceus philippinus*	33
Yellow Wagtail–*Motacilla flava*	35
Brown-headed Barbet–*Megalaima zeylanica*	37
Common Stonechat–*Saxicola torquata*	39
Mynas and Starlings	41

Red-wattled Lapwing–*Vanellus indicus*	43
Yellow-legged Green Pigeon–*Treron phoenicoptera*	45
Rufous Treepie–*Dendrocitta vagabunda*	47
Shikra–*Accipiter badius*	49
Indian Grey Hornbill–*Ocyceros birostris*	51
Rose-ringed Parakeet–*Psittacula krameri*	53
Common Hoopoe–*Upupa epops*	55
Indian Cuckoo–*Cuculus micropterus*	57
Asian Koel–*Eudynamys scolopacea*	59
Common Hawk Cuckoo–*Hierococcyx varius*	61
Black Drongo–*Dicrurus macrocercus*	63
Black Kite–*Milvus migrans*	65
Intermediate Egret–*Mesophoyx intermedia*	67
Rock Pigeon–*Columba livia*	69
Little Brown Dove–*Streptopelia senegalensis*	71
Indian Roller–*Coracias benghalensis*	73
House Crow–*Corvus splendens*	75
Minivets	77
Greater Coucal–*Centropus sinensis*	79
Indian Peafowl–*Pavo cristatus*	81
Shrikes	83
Babblers	85
Tawny Eagle–*Aquila rapax*	87
Grey Francolin–*Francolinus pondicerianus*	89
Common Indian Nightjar–*Caprimulgus asiaticus*	91
Indian Chat–*Cercemola fusca*`	93
Common Iora–*Aegithina tiphia*	95
Grey Tit–*Parus major*	97
Common Chiffchaff–*Phylloscopus collybitta*	99
Common Rosefinch–*Carpodacus erythrinus*	101
CHECKLIST	103
APPENDIX	112

Excerpts from Salim Ali's book
Introduction
The Usefulness of Birds
Bird Watching

DR. KARAN SINGH
MEMBER OF PARLIAMENT
(RAJYA SABHA)
CHAIRMAN
COMMITTEE ON ETHICS

Office : 127, Parliament House Annexe.
New Delhi-110001
Ph. : 2303-4254, 2379-4326
Fax : 23012009
E-mail : karansi@sansad.nic.in

Foreword

Birds are among the most colourful denizens of this planet. The gardens of Delhi, Agra and Jaipur are home to a large number of birds. Shri Samar Singh has done a meticulous study of garden birds which turn out to be as many as 107 species belonging to 33 families, ranging from the glorious Blue Peafowl down to the tiny Minivets and Munias. Shri Samar Singh has with great devotion produced this beautifully illustrated book. This will be a most welcome addition to ornithological literature of the country, and will be particularly welcomed by the large number of bird-lovers. I would like to congratulate Samar Singhji upon the book and hope that it will be widely circulated.

Karan Singh

Nehru's Letter

"If you were with me, I would love to talk to you about this beautiful world of ours, about flowers, trees, birds, animals, stars, mountains, glaciers and all the other beautiful things that surround us in the world. We have all this beauty all around us and yet we, who are grown-ups, often forget about it and lose ourselves in our arguments or in our quarrels. We sit in offices and imagine that we are doing very important work.

"I hope you will be more sensible and open your eyes and ears to this beauty and life that surrounds you. Can you recognise the flowers by their names and the birds by their singing? How easy it is to make friends with them and with everything in Nature, if you go to them affectionately and with friendship. You must have read many fairy tales and stories of long ago. But the world itself is the greatest fairy tale and story of adventure that was ever written. Only we must have eyes to see and ears to hear and a mind that opens out to the life and beauty of the world ..."

3-12-1949
Jawahar Lal Nehru
Former Prime Minister of India

Introduction

Birds have always fascinated humankind and the reasons are quite obvious. Among all the higher forms of life called the vertebrates or back-boned animals, birds are certainly the most beautiful, most melodious, most admired, most studied and most defended. They far outnumber all other vertebrates, except fishes, and can be found virtually everywhere throughout the world. Perhaps the central part of Antarctic is the only place on the world's surface where birds have not been found.

Descended from the reptilian stock similar to the dinosaurs, birds have radiated explosively over the earth in a wide variety of sizes, shapes, colours and habits. Currently, they inhabit every continent and occupy almost every conceivable niche. Some even nest underground. Altogether, there are about 9,000 living species of birds, which the scientists have placed in 27 major groups called Orders and around 155 Families.

Considering that life on earth extended into the spectrum of time for more than two billion years, birds are a latter-day creation. Palaeontologists believe that they began to branch off from the reptilian stock sometime in the late Jurassic period, about 150 million years ago, shortly after the first mammals appeared. Well-known scientist T.H. Huxley described birds as "glorified reptiles" because birds share many characteristics with reptiles, such as certain skeletal and muscular features, somewhat similar eggs and an 'egg tooth' on the upper jaw at hatching time. However, the unique feature that sets them apart from all other life forms is that they have feathers, which are indeed a marvel of natural engineering. No other creatures possess this special feature.

The association between human beings and birds has been very long and intimate. In fact, birds have helped humankind in various ways for thousands of years—from the Geese whose warning cries saved Rome to the Canaries that were used to warn coal miners of methane gas leakage. They continue to provide such lifesaving service by acting as reliable indicators of the health of our environment, specially regarding the dangers arising from chemicals and other toxic substances in the atmosphere. Moreover, birds play a crucial role in maintaining the balance of Nature by controlling insect pests and rodents as well as in cross-pollination of plant species, seed dispersal and as scavengers. Further, there are birds that have made remarkable contributions to human welfare, economically and otherwise. The classic case is that of India's wonder bird, the Red Junglefowl, which is the progenitor of all poultry forms worldwide and has been responsible for several outstanding contributions to medical research and human health.

India's richness in avian diversity is well recognised. Of about 9,000 bird species in the world, around 1,200 are found in India. This means about 13 per cent of the world's total, which is very remarkable for an area that is only about 4 per cent of the world's total landmass. The more spectacular part is the fact that out of 27 orders and 155 families of birds for the world as a whole, India accounts for about 20 Orders and 77 Families. Of the total of 1,200 bird species found in India, about 900 are resident species and the rest, about 300, are migratory, mostly coming from Central Asia and Eastern Europe during the winter period.

Several bird species are quite at home in the urban areas. In this respect, Delhi ranks high as a city with a thriving bird population. In India, it is clearly the richest city in birds, with an amazing list of more than 400 species, which means at least one-third of the country's total number of bird species. Very few cities in the world can boast of such avian richness. Delhi's bird list includes several summer and winter visitors as well as passage migrants, mainly because the city is well positioned along the north-south flowing Yamuna on one of the major Asian flyways. No doubt the most favoured areas are the riverbanks and the Ridge, but several large well-tended parks and gardens and even the trees along the city roads, particularly in south Delhi, attract a rich variety of birds. Agra and Jaipur also have several green spaces that attract birds of different types.

Birds and gardens have a symbiotic relationship for quite obvious reasons. The concept and practice of gardening is certainly homocentric: it is human initiative for human pleasure and benefit. But, in this scheme the natural elements necessarily play a vital role and a garden without trees and birds is inconceivable. Gardens and groves in ancient times were built mostly around temples and palaces. Several sacred groves are still in existence all over the country, but the gardens of the remote past have not survived the vicissitudes of time.

As far as Delhi, Agra and Jaipur are concerned, these were the seats of powerful kingdoms in medieval times and witnessed, under royal patronage, the flowering of human ingenuity in many spheres, including gardening and horticulture. This was specially so during the Mughal period in the 16th and 17th centuries and then again under the British rule in the 19th and 20th centuries. In the post-independence period over the past six decades also several new parks and gardens have been established along with the restoration of those set up earlier. These are like the 'green lungs' of these three important cities, which form the well-known 'Golden Triangle' favoured by the tourists visiting India. An added attraction is, no doubt, the diversity and richness of birdlife harboured in these places.

In this connection, the role of vegetation diversity has to be noted. The more diverse the vegetation, the greater the variety of birdlife therein. This is for the simple reason that each bird species has its preferences and requirements in terms of food, camouflage, roosting, nesting, etc. vis-à-vis trees, shrubs, bushes and other vegetation. Hence, many manicured gardens and parks having only certain types of trees and shrubs generally attract lesser number of birds, whereas the relatively small garden area of the India International Centre in New Delhi, having quite diverse trees and vegetation, accounts for as many as 83 bird species. In short, there is a direct relationship between vegetation diversity and bird diversity.

The list of garden birds in Delhi, Agra and Jaipur

given in the book includes 107 species belonging to 33 different Families. The descriptive notes are on selected species, specially the prominent and interesting ones. As the book title indicates, only those birds have been listed which frequent gardens and parks, including large birds like the glamorous Blue Peafowl, which is our National Bird, as well as tiny birds such as the dainty looking Purple Sunbird, Tailorbird, White-eye and some Minivets and Munias. Care has been taken to include in the list bird species about which there is certainty and to exclude those about whom any doubt has arisen, especially some of the smaller birds whose identification is always difficult. However, suggestions for making additions to the list are welcome.

In a book on birds, it is worthwhile to incorporate some tips on bird watching. Hence, a brief note on the subject by the famous ornithologist, the late Dr Salim Ali, is being included in this publication, along with his views on the usefulness of birds. For giving permission to do so, the Bombay Natural History Society has to be thanked. I also wish to thank Shri P. C. Sen, Director, India International Centre, for entrusting to me this interesting task; Nikhil Devasar and Amano Samarpan, the photographers; and Shobit Arya of Wisdom Tree.

Samar Singh

BIRD DESCRIPTIONS AND PLATES

Little Green Bee-eater (earlier called Small Bee-eater)–*Merops orientalis*
Common Hindi name: Patringa
Size: 18-21 cm, size of Sparrow, but with longer tail
Call: *Tit-tit-tit* uttered repeatedly. Also a shrill sound like *tree-tree-tree*

Dainty looking bird—mostly grass-green, tinged with reddish brown on head and neck, and a rather conspicuous black necklace. Tail prominent and the central pair of tail feathers extend as blunt pins. Bill is long, slender and slightly curved. Male and female look alike.

Generally found in open areas in pairs or parties. Often perches on fence posts and transmission lines. Large numbers collect before roosting time in evening and indulge in flying about and twittering. Common mainly during summer.

Feeds mainly on insects and relishes bees. Nesting time is in summer and the nest is made like a tunnel in the side of a borrow-pit or in sandy ground. Both parents share in excavating the nest and raising the chicks.

The Blue-cheeked Bee-eater (*Merops persicus*) is larger in size and has a conspicuous white and blue-green patch on the cheek.

Little Green Bee-eater – *Merops orientalis*

Oriental White-eye–*Zosterops palpebrosus*
Common Hindi name: Baboona
Size: About 10 cm, smaller than Sparrow
Call: Soft twittering. Breeding male is more noisy and somewhat musical

Tiny bird with a square tail, greenish yellow plumage and a conspicuous white ring around each eye. Bill is slender, pointed and slightly curved.

Gregarious by nature and generally found in small parties. Moves energetically in the foliage of bushes, shrubs and trees and subsists mostly on the nectar of flowers. Has a role in cross-pollination. Also feeds on berries and small insects.

With the onset of spring, these birds become really active and are very noticeable. This coincides with the breeding season, which extends up to July-August. The nest is neatly woven and cup-shaped, which is placed hammock-like in the fork of a twig above the ground. Both sexes share the duties of raising the chicks.

Oriental White-eye – *Zosterops palpebrosus*

Purple Sunbird–*Nectarinia asiatica*
Common Hindi name: Phool Soongni
Size: About 10 cm, smaller than Sparrow
Call: Sharp notes that sound like *wich-wich-wich* made while flitting about. Breeding male utters *cheewit-cheewit* repeatedly, while raising and lowering the wings

Tiny, attractive bird. Adult male's breeding plumage is metallic dark blue and purple, which looks black from a distance and glistens in sunlight. Also has brilliant yellow and scarlet tufts of feathers on the underside of the wings. Non-breeding male and female look alike in being olive-brown above, pale yellow below and with darker wings and a broad black stripe running down the breast.

Both sexes have a slender, pointed and curved bill and a tubular tongue very suited to probing flowers and sucking the nectar, which forms the main diet of the bird and explains its Latin name. In playing this role, the bird aids in cross-pollination. Also feeds on small insects, including spiders.

Generally moves in pairs and frequents gardens, groves and open forest areas. Nesting time is during the summer months. Female makes the nest and incubates; male helps in feeding the chicks.

Purple Sunbird – *Nectarinia asiatica*

Blue Rock Thrush–*Manticola solitarius*
Common Hindi name: Pandoo
Size: 22-25 cm, size of Bulbul
Call: Mostly silent, though the male does make some whistling sounds

Breeds during summer in the Himalaya region and moves in winter to the plains. Occasionally seen at this time in parks and gardens.

Male is bright indigo blue, with blackish wings and tail. Female is mostly grey-brown above, whitish below cross-barred with dark brown, and with a pale wing-bar.

Feeds mostly on insects and also berries.

Blue Rock Thrush – Manticola solitarius

Oriental Magpie Robin–*Copsychus saularis*
Common Hindi name: Dhaiyal or Dhaiyar
Size: 20-22 cm, size of Bulbul
Call: More vocal in breeding season; has several pleasant notes. Can also mimic other birds

Familiar bird in and around human habitations. Male has a black-and-white plumage, while female is more slaty grey and brownish. The white-fringed tail is prominent and remains cocked. Only when calling, the tail is jerked up and down.

Generally found in pairs. Becomes active and noisy during the summer months, coinciding with the breeding season. In winter, the bird becomes shy and silent.

Feeds on small insects and also the nectar of some flowers, thereby aiding cross-pollination.

Oriental Magpie Robin – *Copsychus saularis*

Asian Paradise Flycatcher–*Terpsiphone paradisi*
Common Hindi name: Dudhraj/Shah Bulbul
Size: 20-22 cm, size of Bulbul, but with long tail
Call: Harsh grating *che* or *che-chew*. Also some nice notes during breeding season

The silvery white adult male, with two long ribbon-like feathers in tail, is a star attraction. Young male and female are chestnut above and greyish-white below. Head is metallic black and has a prominent crest as in the Bulbuls. Mostly sighted in pairs, flitting from tree to tree. The quick movements of the white male, with the long tail feathers trailing behind, are a spectacle indeed.

Feeds mostly on insects. Nesting time is during the summer season and the cup-shaped nest is generally made in the crotch of a twig well above the ground. Female plays the major role in rearing the chicks.

The most likely period for sighting these birds is the spring season (February to early April). This is also the time when the gorgeous plumage of the male is at its best.

Asian Paradise Flycatcher – Terpsiphone paradisi

House Swift–*Apus affinis*
Common Hindi name: Babeela
Size: Around 15 cm, size of Sparrow
Call: Shrill twittering screams, specially in evenings

Small, smoky-black bird, with white throat and rump, short, square tail and long narrow wings. Male and female are, more or less, alike.

Prefers old buildings, monuments and ruins. Usually flies about at high speed, zipping in and out almost incessantly, while effortlessly hawking flies and small insects. A thick fringe of eyelash-like feathers protect the eyes from accidental mid-air collision with objects. More active and noisy during the evening hours.

The foot structure of the bird is such that it cannot perch normally and hence merely clings to rough surfaces. During the nesting season, from February to September, nest colonies of these birds can be seen on sheer vertical surfaces, including walls and ceilings. Both sexes share in rearing the chicks.

Quite common, except during winter. Evidently, local migrant to Delhi.

House Swift – Apus affinis

Red-vented Bulbul–*Pycnonotus cafer*
Common Hindi name: Bulbul
Size: About 20 cm, slightly smaller than Myna
Call: Chattering notes, sometimes musical

Smoky-brown in colour, with scale-like markings on back and breast, white rump and a conspicuous crimson patch below root of tail. Head is partially crested and black. Male and female look alike.

Frequents gardens, parks and light scrub areas and also enters homes to feed on kitchen scraps. Perky and confiding by nature. Feeds on insects, berries, fruits and flower nectar. Nesting takes place between February and May and the nest is made in a bush or tree. Both sexes share parental duties.

The Red-whiskered Bulbul *(Pycnonotus jocosus)* is also seen. It is the same size as the Red-vented Bulbul and has almost the same habits. But, its plumage is more eye-catching, as it has a fuller black crest and rather distinctive crimson whiskers, apart from the crimson patch under the root of tail, as in the Red-vented Bulbul.

Yet another interesting species in the group is the White-cheeked Bulbul (*Pycnonotus leucotis*) whose distinctive features are the glistening white cheeks and the yellow vent (under the tail).

Red-vented Bulbul – *Pycnonotus cafer*

White-throated Kingfisher (earlier called White-breasted Kingfisher)–*Halcyon smyrnensis*
Common Hindi name: Kilkila
Size: Around 28 cm, smaller than Pigeon
Call: Shrill *kilikili* type of territorial call, usually from a tree perch, and also a loud cackling call when in flight

Usually seen near water bodies, but also found in parks, gardens and cultivated fields. More common than other Kingfishers.

Has brilliant turquoise-blue plumage with deep chocolate-brown head, neck and underparts, a rather conspicuous white 'shirt front' and a long, heavy, pointed bill. A white wing-patch is prominent during flight. Both sexes look alike.

Feeds on small fish, tadpoles, lizards and insects. Occasionally takes mice and young birds. Nesting season is during the summer months, usually in tunnels excavated in earth-banks near water. Both parents raise the chicks.

The Pied Kingfisher (*Ceryle rudis*) is also an interesting bird with different plumage but with somewhat similar habits as the above-metioned Kingfishers.

White-throated Kingfisher – *Halcyon smyrnensis*

Common Kingfisher (earlier called Small Blue Kingfisher)–*Alcedo atthis*
Common Hindi name: Chhota Kilkila
Size: 16-18 cm, larger than Sparrow
Call: High-pitched shrill *chee-chee* or *chit-it-it*, repeated rapidly

Very attractive small bird with a shimmering sapphire-blue head, back and body, contrasted by a fiery rust-orange breast and belly. These bright colours are very noticeable even during flight. Bill is long, straight and pointed like a dagger, which is used deftly in picking up the prey. Tail is stumpy; legs are short and orangish. A white patch behind the eye and nape, and under the chin, is quite distinctive. Both sexes look almost alike, although the female has orange colouration on the bill's lower part.

Generally feeds on small fishes, tadpoles and aquatic insects. Nesting is during the summer months, usually in tunnels excavated in earth-banks near to water. Both sexes raise the chicks.

Common near ponds and small water bodies.

Common Kingfisher – Alcedo atthis

Spotted Owlet–*Athene brama*
Common Hindi name: Oolloo
Size: 20-22 cm, size of Myna
Call: Several harsh chattering notes, often heard at night, sometimes simultaneously by two individuals

Mainly nocturnal. Daytime is usually spent in some hollow inside a tree-trunk or huddled on a secluded branch.

Small and squat in body, which is spotted white and greyish-brown. The head is disproportionately large and round, with somewhat staring yellowish eyes, which impart a rather clownish appearance.

Feeds on beetles and other insects and also young birds, mice and lizards. Nesting is done mostly in a tree hollow or in a hole in some old unoccupied building. Both sexes share the domestic duties.

Spotted Owlet – *Athene brama*

Common Tailorbird–*Orthotomus sutorius*
Common Hindi name: Darzee/Phutki
Size: 12-13 cm, slightly smaller than Sparrow
Call: Sounds like *towit-towit-towit* uttered quite loudly and repeatedly

Small bird, mostly olive-green, with whitish underparts, rust coloured crown and two pin-pointed and cocked tail feathers. Male and female look alike. Generally found in pairs.

Fond of shrubbery and frequents gardens, parks and even homes, where it moves swiftly amongst creepers and other plants looking for flower nectar or small insects.

Nesting time extends from April to September. The nest is made neatly like a small funnel by folding and fashioning one or more plant leaves. Both sexes share the domestic duties.

Common Tailorbird – *Orthotomus sutorius*

Yellow-crowned Woodpecker (earlier called Yellow-fronted Pied Woodpecker)– *Dendrocopos mahrattensis*
Common Hindi name: Katphora
Size: 17-18 cm, size of Bulbul
Call: Sharp *click, click* or *click-r-r-r*

Small bird with long, pointed, stout bill and wedge-shaped tail. Irregularly spotted black and white above. Whitish, brown-streaked breast and flanks; scarlet-crimson on abdomen and vent. Female without scarlet in crest. Frequents open scrub country, gardens and groves with trees. Moves singly or in pairs. Generally clings to tree-trunks and keeps tapping on the bark and digging into rotten wood for insects and grubs. Scuttles up the tree-trunk in jerky spurts using the tail as the third leg to cling and balance. When in flight, the wings are beaten rapidly for some time and then closed for a pause, quite typical of Woodpeckers.

Nesting time is from January to May. The nest is made as a hole inside a decaying tree branch at some height. Both sexes share the domestic duties.

The larger Black-rumped Flameback (*Dinopium benghalensis*) was earlier called the Lesser Golden-backed Woodpecker. The male has a golden yellow and black plumage along with a distinctive crimson crown and crest. In habits, it is much like the Yellow-crowned Woodpecker.

Yellow-crowned Woodpecker – *Dendrocopos mahrattensis*

House Sparrow–*Passer domesticus*
Common Hindi name: Gauraiya
Size: about 15 cm
Call: Quick chirping notes like *chi, chi, chi* and *chir, chir, chir*, uttered repeatedly

Commonest and most familiar bird. Always found in and around human habitations and in gardens, agriculture fields and other open areas. Highly gregarious and generally moves in flocks or small parties. Believes in community roosting, preceded by much noisy chirping.

Male is quite attractive with a black bib, ashy grey crown and black-streaked chestnut plumage. Female is brownish and streaky looking.

Omnivorous and will eat almost anything from grain and kitchen leftovers to flower nectar and small insects. Nesting is practically round the year.

In recent years, there has been some decline in numbers.

About the same size as the House Sparrow is the Chestnut-shouldered Petronia or the Yellow-throated Petronia (*Petronia xanthocollis*). As the name suggests, it has a conspicuous chestnut colour patch on the shoulder and a yellow patch on the throat and white bars on the wings, but overall the bird is earthy brown. The female lacks the yellow throat-patch and is more pale in plumage.

House Sparrow – *Passer domesticus*

Indian Robin–*Saxicoloides fulicata*
Common Hindi name: Kalchuri
Size: 16-17 cm, size of Sparrow
Call: No distinct call, though the male utters some shrill notes, especially in the breeding season

Very familiar bird. Male is mostly black, with a white patch on the wings and rusty red below the cocked tail. Female is mostly ashy brown.

Generally moves in pairs and is quite confiding, hence freely seen in human habitations. Hops on ground, feeding on insects and grain.

Nesting time is in the hot season and the nest is usually made on the ground under a stone or in a hole in earth-bank or in a tree-stump. Both sexes take part in raising the chicks.

Indian Robin – *Saxicoloides fulicata*

Golden Oriole–*Oriolus oriolus kundoo*
Common Hindi name: Peelak
Size: About 25 cm, little bigger than Myna
Call: Harsh notes that sound like *cheeah-cheeah* and also fluty whistles like *pee-lo-lo, pee-lo-lo*

Male is strikingly beautiful—mostly bright golden yellow, with black in wings and tail and a conspicuous black streak running through each eye. Female is duller and greener. Bill is quite prominent and yellowish. Despite the brilliant plumage, the bird is not always easy to spot because its pattern of yellow and black somehow merges with the sunlight and shadows and the foliage through which the bird flits about.

Arboreal by nature and partial to wooded areas, groves and gardens with large trees. Generally seen in pairs.

Nesting is during the summer months. The nest is cup-shaped, neatly woven and suspended hammock-like in the fork of a leafy twig on a tree. Both parents take part in raising the chicks.

Golden Oriole – Oriolus oriolus kundoo

Baya Weaver–*Ploceus philippinus*
Common Hindi name: Baya/Son Chiri
Size: 15-16 cm, size of Sparrow
Call: Sparrow-like notes, mostly *chit, chit*. Breeding males make joyous *chee-ee, chee-ee* notes in chorus

Breeding male has lovely bright yellow crown and dark brown upperparts, streaked with yellow. Breast is also yellow and underparts are creamish buff. Non-breeding male and female look like the female of House Sparrow—streaked brown above and pale white below. Usually operate in small parties or flocks and feed mainly on grass, seeds, grains and tiny insects.

The most interesting feature of the species is its remarkable skill in weaving colony nests. Each nest is an architectural marvel. It is made like an upside down flask, with a vertical tubular entrance and a compact egg chamber inside. Strips of grasses and paddy leaves are mostly used for the purpose. These are woven skilfully with the help of the strong conical bill to make a neat nest, which is hung deftly on a branch, mostly over water, such as dug wells. The whole operation extends over some days and involves several hundred sorties to gather the materials. Another interesting aspect is that only the males build the nests and each male makes several nests and has as many females at the same time. However, the females are responsible for incubation and raising the chicks.

Baya Weaver – *Ploceus philippinus*

Yellow Wagtail–*Motacilla flava*
Common Hindi name: Pilkiya
Size: 17-18 cm, size of Sparrow
Call: Loud, distinct notes that sound like *weesp* and *pseeu, pseeu,* uttered at intervals

Slim and long-tailed; yellowish or olive-green above and yellow below. Lively by nature and can be seen running about on ground in short spurts, wagging its tail up and down and picking up small insects, which form the main diet. Flies in undulating curves.

Breeds during summer in the western Himalaya region and then in winter, large flocks, literally in swarms, descend to the plains. Quite common then.

The Grey Wagtail *(Motacilla cinerea)*, Citrine Wagtail *(Motacilla citreola)* and White Wagtail *(Motacilla alba)* are also seen occasionally during winter. In build and habits, they are all quite similar, but there are slight differences in plumage.

Yellow Wagtail - *Motacilla flava*

Brown-headed Barbet (earlier called Large Green Barbet)–*Megalaima zeylanica*
Common Hindi name: Bada Basanta
Size: 27-28 cm, size of Myna
Call: Sounds like *kor-r-r-kutroo, kutroo, kutroo,* that go on and on specially during summer, often in chorus with others of the species

The most common Barbet, stout and heavy-billed, with head, neck, upper back and breast mostly brown and having off-white streaks. Lower breast and abdomen grass-green; under-tail coverts bluish and a conspicuous orange patch around each eye that extends to the base of bill.

Arboreal by nature and rather difficult to spot because its plumage merges well with the green foliage of trees. Feeds mostly on figs, berries, fruits, and flower nectar; sometimes takes insects as well. Nesting season is from February to June. The nest is made by excavating a hole in a tree branch well above the ground. Both sexes share the domestic duties.

Another Barbet to be seen is the smaller Coppersmith Barbet (earlier called Crimson-breasted Barbet). It is a cute bird with crimson forehead and breast, yellow throat and green streaked yellowish underparts. In habits, it is much like the Brown-headed Barbet, except for a typical *tok-tok-tok* call in long runs.

Brown-headed Barbet – *Megalaima zeylanica*

Common Stonechat (also called Collared Bushchat)–*Saxicola torquata*
Common Hindi name: Khar Pidda
Size: 14-15 cm, size of House Sparrow
Call: Chattering notes such as *chek, chek* and *pee-tack, pee-tack*

Small attractive bird, specially the male, which is peppercorn black above and rufous ochre below, with a prominent white collar and also a white streak on each wing. The female is mostly brown, dark streaked above and rufous below.

Essentially a winter visitor, as it breeds in the Himalayan region and arrives around September, after which it stays on till April. Usually seen singly or in pairs, usually perched on low bushes, boulders or rocks or tall grass stems. Feeds mostly on small insects. A typical habit is the constant up-and-down flicking of the tail.

The slightly smaller and dumpier black-and-white Pied Bushchat (*Saxicola caprata*) is also seen. It looks somewhat like the Magpie Robin, but the habits are much like those of the Common Stonechat. However, it is a resident bird.

Common Stonechat – *Saxicola torquata*

Mynas and Starlings

Mynas and Starlings belong to the same family, Sturnidae, but are described separately. Sometimes this creates confusion. In India, eleven species are known, of which five are Mynas and six are Starlings. The following are frequently seen in gardens:

> Common Myna *(Acridotheres tristis), below left*
> Bank Myna *(Acridotheres ginginianus), above right*
> Common Starling *(Sturnus vulgaris), below*
> Brahminy Starling *(Sturnus pagodarum), below right*
> Asian Pied Starling *(Sturnus contra), above left*

These birds have certain differences in plumage, but in habits there are many similarities. In particular, they are all perky and gregarious, confiding by nature and frequent gardens and homes looking for something to feed on. Besides, they keep chattering a great deal during daytime and spend considerable time on the ground. They are mostly omnivorous.

Mynas and Starlings

Red-wattled Lapwing–*Vanellus indicus*
Common Hindi name: Titehri or Titori
Size: 32-35 cm, size of Partridge but with longer legs
Call: Loud and penetrating, somewhat like *'did-he-do-it'*, *'did-he-do-it'*

The more common plover. Upperparts bronze-brown in colour, while the underparts are white. Head, neck and breast are black. A distinctive fleshy reddish wattle exists in front of each eye and a broad white band runs from behind the eye to the neck and the underparts. The legs are long and yellow. Male and female look identical.

Prefers open dry country, ploughed fields and also parks and gardens. Usually found in pairs. Runs about in short spurts and is always alert and vigilant. Also the first to raise an alarm at the slightest provocation.

Feeds on insects, molluscs and grain. Nesting time is during the summer. Eggs are laid on bare ground and match the colour of soil. Besides, they are skilfully camouflaged to prevent detection. The parents remain ever vigilant and raise alarm with their screams on any intrusion and may even attack the intruder.

Red-wattled Lapwing – *Vanellus indicus*

Yellow-legged Green Pigeon–*Treron phoenicoptera*
Common Hindi name: Harial
Size: 32-34 cm, like Pigeon
Call: Typical musical whistles, quite pleasing, often heard in morning hours

Stockily built but pretty looking bird. Yellow, olive-green and ashy grey in colour, with a lilac patch on shoulders and a yellow bar in the blackish wings. Legs are distinctly yellow. Both sexes look alike.

Moves in flocks and flies swiftly. Rarely seen on the ground. Prefers fig-bearing trees, specially ficus species. Feeds mostly on figs, berries and other fruits. The ficus trees attract these birds in large numbers, specially from February to April.

Nesting time is in the summer months. Both parents take care of the chicks.

Yellow-legged Green Pigeon – *Treron phoenicoptera*

Rufous Treepie (earlier called Indian Treepie)– *Dendrocitta vagabunda*
Common Hindi name: Mahalat/Kokila
Size: 46-50 cm, bigger than Myna and has much longer tail
Call: Large repertoire, some harsh and loud, others melodious like *Kokila* and *bo-bo-link*

Handsome bird with a long black-tipped tail (about 30 cm). Head and neck are sooty, while the body is mostly chestnut-brown and wings have grey and white patches. Bill is stout like that of a Crow. Also known as Indian Treepie.

Arboreal by nature and prefers wooded areas, parks and gardens with plenty of trees. Moves in pairs or small parties and makes presence known by noisy calls. Has wandering and somewhat vagrant habits, which explains its second Latin name!

Omnivorous and takes fruits, berries, insects, frogs, rodents, lizards and even young birds. Nesting time extends from February to July and the nest is made on a tree. Both parents share in raising the chicks.

Rufous Treepie – Dendrocitta vagabunda

Shikra–*Accipiter badius*
Common Hindi name: Shikra
Size: 30-36 cm, bigger than Pigeon
Call: Loud, harsh and like *ti-tu titu* or *ke-kee ke-kee*

A skilled raptor (bird of prey), which generally waits in ambush in wooded cover and dives swiftly after its target—small birds, animals, lizards, etc. Usually flies close to the ground, then shoots up to alight on a branch. Most prey is taken from the ground, but some caught on the wing. Also takes young birds out of nests.

Hawk-like and lightly built. Ashy blue grey above and white below, cross-barred with rusty brown. The female is more brownish above and larger than the male. Legs are yellow. Tail has broad blackish cross-bands and a dark tip. Bill is small and fiercely hooked, typical of a Hawk. Eyes have orange rings.

Nesting season is during the summer months. Female incubates; male also takes part in raising the chicks.

Shikra – *Accipiter badius*

Indian Grey Hornbill–*Ocyceros birostris*
Common Hindi name: Dhanesh
Size: 55-60 cm, size of Black Kite
Call: Cackles, like *k-k-k-kae* and also chattering notes, like *cheu-cheu-cheu*

Plumage is brownish grey; the tail is long, graduated and white-tipped. The black-and-white curved bill (about 3" long) with a peculiar bony casque is a prominent feature. The bill can hold several berries at a time, which helps in feeding the female and chicks during nesting season.

The nest is usually a deep cavity in a hollow tree trunk, with an opening like a front door entry. The female uses the pre-incubation period for laying eggs (2-3) and walling up the nest with her droppings, lumps of mud brought by the male and even wood chips. The entrance is sealed in the same way and the female stays inside for the next 75 days or so, being fed from outside through a narrow slit by the male. The entry point is opened after the chicks are hatched, after which both parents take part in rearing the young. This very fascinating practice surely provides protection during the most vulnerable period, apart from being a safe way to ensure procreation.

Essentially arboreal in nature, with preference for open wooded country, large groves and gardens. Feeds mostly on fruits and berries; also large insects, small mice and lizards. Gardens with several old and big trees, especially the ficus species, attract these birds, usually in pairs and small parties, round the year. More noticeable from around February.

Indian Grey Hornbill – *Ocyceros birostris*

Rose-ringed Parakeet–*Psittacula krameri*
Common Hindi name: Tota
Size: 40-42 cm, size of Myna, but with a long, pointed tail
Call: Loud sharp screams like *keeak, keeak, keeak*

Most common and abundant parakeet in the country. Has also become a popular cage bird and is good at mimicking voices, including human ones. Has a pleasing grass-green plumage with a long, pointed tail and a large, hooked red bill. Male also has a prominent black and rose-pink collar, which is missing in the female.

Noisy bird, especially in large flocks that are common. Flies swiftly, covering long distances. Feeds mostly on fruits and berries; also takes other vegetable matter. Nesting time is mainly during the spring season. The nest is usually made in a tree-trunk hollow. Both parents share the domestic duties.

The slightly larger Alexandrine Parakeet (*Psittacula eupatria)* and the smaller Plum-headed Parakeet (*Psittacula cyanocephala*) are also very attractive birds and have their own distinguishing features.

Rose-ringed Parakeet – *Psittacula krameri*

Common Hoopoe (earlier called Hoopoe)–*Upupa epops*
Common Hindi name: Hudhud
Size: 30-32 cm, size of Myna
Call: Soft but penetrating *hoo-po* or *hoo-po-po*, uttered repeatedly, sometimes for 5-10 minutes at a stretch

Fawn coloured, with prominent black-and-white zebra-like markings on back, wings and tail, a conspicuous fan-shaped crest, and long, slender, gently curved bill. Male and female look alike.

Usually moves in pairs and prefers open areas, gardens, groves and parks. On the ground, it has a peculiar waddling gait. Feeds mostly on tiny insects, picked on the ground.

Nesting takes place during the summer months and the nest is generally made in a tree-hollow or a hole in wall or ceiling of a building. It is made untidily and kept quite unclean. Both sexes share in rearing the chicks.

Common Hoopoe – Upupa epops

Indian Cuckoo–*Cuculus micropterus*
Common Hindi name: Cuckoo
Size: 32-34 cm, size of Pigeon but with longer tail
Call: Distinct loud fluty notes that sound like *orange-pekoe, kaiphal-pakkal-bo-kotako,* etc. Calls are repeated, often endlessly

Winter visitor. Breeds during summer in the Himalaya region.

Dark slaty grey in colour, with a tinge of brown above. Pale ashy and white below, barred with broad brownish bands. There is a broad black sub-terminal band on the tail, which is special to this Cuckoo. Female is more brown and has a distinctive brown patch around each eye.

Prefers to stay in trees, often singly. Feeds mostly on caterpillars and other insects. Nesting season is during the hot weather, but the bird does not make its own nest and lays eggs in the nest of some other bird and leaves it to the foster parents to incubate and bring up the chicks. This is 'brood parasitism', a typical trait of most Cuckoos.

Indian Cuckoo – *Cuculus micropterus*

Asian Koel (earlier called Common Koel)– *Eudynamys scolopacea*
Common Hindi name: Koel
Size: 41-44 cm, size of House Crow, but with longer tail
Call: Unmistakable and quite familiar *kuoo-kuoo-kuoo,* that builds to a crescendo, usually with the start of summer and extends through the monsoon period. Generally silent in winter

Summer visitor in Delhi. Male is glistening black, with yellowish green bill and crimson eyes. Female is brown, profusely spotted and barred with white (*facing page*).

Prefers large leafy trees and hence seen often in large groves and gardens, singly or in pairs. Feeds mostly on fruits and berries, but also picks up insects.

Nesting season is during the summer months, but the bird does not believe in making its own nest. Instead, the female slyly lays eggs generally in a Crow nest and virtually leaves it to the foster parents to hatch and rear the chicks—a special trait of most Cuckoos.

Asian Koel – *Eudynamys scolopacea*

Common Hawk Cuckoo (earlier called Brainfever Bird)–*Hierococcyx varius*
Common Hindi name: Papiya or Papiha
Size: 32-34 cm, size of Pigeon, but with a longer tail
Call: Loud screams that sound like *brain-fever, brain-fever*, repeated 5-6 times and ending abruptly, and then again and again, virtually through the day and often during moonlit nights. Calling intense during summer and through the rainy season. In winter, mostly silent

Ashy grey above, whitish below, cross-barred with brown. The long tail is broadly barred. Looks somewhat like the Shikra, but more slender. Frequents groves and gardens with trees, especially fruit trees. Feeds on berries, figs and fruits and also on caterpillars and insects.

Nesting time is during the summer months. Like most Cuckoos, this bird practices brood parasitism, i.e. lays eggs in the nest of another bird and leaves it to the foster parents to hatch and rear the chicks.

Common Hawk Cuckoo – *Hierococcyx varius*

Black Drongo–*Dicrurus macrocercus*
Common Hindi name: Kotwal/Bhujanga
Size: 28-32 cm, size of Bulbul
Call: Some notes are harsh and challenging and others fluty. Also capable of mimicking other birds

Glossy black bird throughout, with a distinctive long and forked tail. Male and female look alike. Generally seen in open country, often perched on electric or telephone lines or even in the company of cattle, riding on their backs and picking up insects. The bird has a knack of getting to spots where insects abound and these form its main diet. For this reason, it is considered very beneficial to agriculture.

Nesting is generally from April to August. The nest is flimsy, cup-shaped and made in the fork of a tree branch. Both parents display considerable boldness in protecting the nest and the chicks.

The Ashy Drongo (*Dicrurus leucophaeus*) may also be sighted occasionally during winter. It is like the Black Drongo except for the slaty colouration above and dull grey below. It breeds during summer in the Himalaya region and is a winter visitor to the plains. Yet another species is the White-bellied Drongo (*Dicrurus caerulescens*). As the name indicates, its white belly is distinctive, but it is not so common.

Black Drongo – *Dicrurus macrocercus*

Black Kite (earlier called Pariah Kite)–*Milvus migrans*
Common Hindi name: Cheel
Size: 55-68 cm, smaller than Vulture
Call: Shrill, almost musical, whistling *ewe-wir-wir-wir,* uttered while perched or on wing

Belongs to the Hawk family and essentially a bird of prey (also called raptor), which means that it specialises in the act of preying on other life forms, mainly small birds, animals, fishes, amphibians, reptiles, insects, etc. Also plays the role of an efficient scavenger of animal remains and carcasses. Gregarious and very adaptable, hence successful in nature. Numerous mainly around habitations.

From a distance, the bird appears almost black, but actually its plumage is dark brown, with scattered light brown rufous markings, particularly on the head, neck and underparts. The most distinguishing feature is the forked tail, which is clearly noticeable in flight. Can perform remarkable manoeuvres while flying fast and also capable of merely gliding and hovering.

Nesting season is from around September to April; nest is usually made high on a large tree. Both parents share in raising the chicks.

Two common kites found in parks and gardens are the Black-shouldered Kite (*Elanus caeruleus*) and Brahminy Kite (*Haliastur indus*). Both are smaller than the Black Kite.

Black Kite – *Milvus migrans*

Intermediate Egret (earlier called Median Egret)–
Mesophoyx intermedia
Common Hindi name: Bagla
Size: Around 65 cm, bigger than Chicken
Call: Some indistinct croaking

Common near water bodies. Called Intermediate or Median to distinguish from the Large Egret and the Little Egret, apart from the Cattle Egret. Lanky and pure white, with a long neck and a large yellow bill. Has a small crest, which disappears in the breeding season (monsoon period), when the yellow bill turns black. In fact, the breeding plumage of the bird is quite distinctive in terms of the filamentous plumes that appear on the back and the breast at that time. The long legs, typical of wading birds, are black.

Feeds mostly on fishes, frogs and insects that are stalked on mud, grass or shallow water. Roosts at night in favourite trees and the nests are also made on such trees. Both sexes take part in raising the chicks.

Intermediate Egret – Mesophoyx intermedia

Rock Pigeon (earlier called Blue Rock Pigeon)–
Columba livia
Common Hindi name: Kabutar
Size: 32-34 cm, smaller than House Crow
Call: Familiar *gootr-goo, gootr-goo*, uttered repeatedly

Slaty grey bird with glistening metallic green, purple and magenta sheen on neck and upper breast. Two dark bars appear on each wing and a band runs across the tail-end. Both sexes look identical.

Usually seen in flocks. The wild variety prefers open country with cliffs and rocky hills. The other is well adapted to human habitations, even densely populated urban areas, and virtually semi-domesticated. Essentially a vegetarian, feeding on seeds and grains. Nesting season is not defined. Eggs are laid in a small nest that may be made on a ledge or even the ceiling of a dwelling house. Both sexes share in the domestic duties.

Rock Pigeon – *Columba livia*

Little Brown Dove (also called Laughing Dove)–
Streptopelia senegalensis
Common Hindi name: Panduk/Fakhta
Size: 26-28 cm, slightly bigger than Myna
Call: *Coo-rooroo-rooroo* uttered softly

Slim bird with a prominent well-formed tail. Colour is earthy brown and grey above, pinkish brown and white below with a miniature chess-board in rufous and black on either side of neck. Both sexes look alike. Male puts on a typical display during courting, when it starts bobbing and calling at the female and also advancing in stiff hops on the ground.

Usually seen in pairs or more numbers. Prefers dry scrub country, but also at home inside habitations and homes. By nature, tame and confiding. Feeds on seeds and grains. Nesting season is almost through the year. Nest can be in any safe place. Both sexes take part in raising the chicks.

Other doves noticed in gardens are the Spotted Dove (*Streptopelia chinensis*) and the Oriental Turtle Dove (*Streptopelia orientalis*). The first is pinkish brown and spotted, while the second is more reddish brown Both are bigger than the Little Brown Dove.

Little Brown Dove – *Streptopelia senegalensis*

Indian Roller–*Coracias benghalensis*
Common Hindi name: Nilkanth
Size: 32-34 cm, size of Pigeon
Call: Loud and raucous croaks and chuckles, including one that sounds like *chack, chack*

Colourful bird–crown, wings and tail in varying shades of blue, while the breast and some portion of upperparts are rufous brown. Head is large and bill heavy, slightly curved and black. Male and female look alike.

Usually seen singly or in pairs, often perched on transmission lines or fence posts. Prefers open areas, parks and gardens. Feeds on insects, frogs and lizards. Its propensity to devour large quantities of insects is considered beneficial to agriculture.

Male indulges in courtship display in the breeding season, which is during early part of summer. Nest is usually made in a tree-hollow and sometimes in a hole in the wall of a building.

Some people consider the sighting of this bird as auspicious. In some erstwhile princely states, it was customary to release the bird on special occasions, such as the Dussehra festival.

Indian Roller – *Coracias benghalensis*

House Crow–*Corvus splendens*
Common Hindi name: Kowwa/Kagla
Size: 40-43 cm
Call: Familiar *kaw-kaw-kaw*

Among the more common birds and virtually found everywhere. Highly adaptable and its numbers seem to be increasing. Make-up, habits and commensal relationship with humankind are well known.

Its role as a scavenger in nature has assumed added importance following the decline in the population of Vultures. At the same time, the increase in Crow population is posing problems for other birds, whose eggs and young ones are often pilfered by Crows.

Sometimes, the jet-black, large-billed Jungle Crow is also seen, though in very small numbers and usually in the company of House Crows. The House Crow is smaller in size and has a distinctly grey neck.

House Crow – *Corvus splendens*

Minivets

Minivets are a group of brightly coloured birds. By habit, they are arboreal, gregarious and mainly insectivorous. Three more commonly sighted species are:

   Small Minivet *(Pericrocotus cinnamomeus), below left*
   Scarlet Minivet *(Pericrocotus flammeus), below right*
   Long-tailed Minivet *(Pericrocotus ethologus), top*

   The Small Minivet is more common; it is smaller than a Sparrow and has lovely grey, black and orange-crimson plumage. The Scarlet Minivet is larger and its male is very attractive in glistening black and orange-red to deep scarlet. Likewise, the Long-tailed Minivet is glossy black and scarlet all over. It is essentially a winter visitor, as it breeds during summer in the Himalayan region, like the Rosy Minivet.
   Another species in the group is the White-bellied Minivet (*Pericrocotus erythropygius*), which is less arboreal than other minivets.

Minivets

Greater Coucal (also called Crow Pheasant)– *Centropus sinensis*
Common Hindi name: Coucal/Mahoka
Size: 46-48 cm, bigger than House Crow and with a long broad tail
Call: Typical *coop-coop-coop*, deep and resonant, repeated quickly in a tempo. Also utters some harsh croaks and gurgling chuckles

Also a Cuckoo, but not parasitic. In other words, it makes its own nest, mostly during summer, but sometimes extending into the rainy season. The nest is generally made in a thorny shrub and both parents take care of the chicks.

Largely terrestrial and prefers areas with plenty of shrubs and undergrowth. Shuffles about on the ground and then clambers and hops with agility on the branches of trees and shrubs. Feeds mostly on caterpillars, insects, small mice and even bird eggs.

In appearance, the bird is glossy black, with rather conspicuous chestnut wings and a long, broad, graduated tail. Eyes are crimson, bulging and have eyelashes. Male and female look identical.

Greater Coucal - Centropus sinensis

Indian Peafowl (also called Blue Peafowl)–
Pavo cristatus
Common Hindi name: Mor/Mayur
Size: *Male*: 180-230 cm in full plumage. *Female*: 90-100 cm
Call: Loud and distinct *may-awe, may-awe*, invariably heard early in the morning. This call gains in frequency with the arrival of monsoon. Also utters short, ringing shrieks like *ka-ann, ka-ann*, with pumping action of head and neck, mostly to signal alarm

The National Bird of India. Has received special attention in Indian mythology, religion, literature, art and music down the ages.

The plumage of the adult male, along with a prominent crest and glossy green train of elongated upper tail-covert feathers (1 to 1.5 m in length) having numerous ocilli, is spectacular indeed. The Peacock lifts these feathers in a fascinating fan-shaped pattern for the well-known dance, performed mostly to attract the attention of the females, who are far less attractive. The male is polygamous and usually has 4-5 females in company.

Very shy and ever alert by nature, but very adaptable and quite at home near human habitations, including urban areas having undergrowth and tree cover. Mostly stays on ground but roosts on large trees at night. Nesting time extends from January to October. The nest is made on the ground. Female incubates and plays the major role in raising the chicks.

Indian Peafowl – *Pavo cristatus*

Shrikes

Shrikes are a large group of birds. In size, they are between Bulbul and Myna. The distinctive features are large heads, strong hooked bills and sharp Hawk-like claws. They are known as 'butcher birds' on account of the habit of killing more than they can consume at a time and saving the surplus for later. They are good at mimicking and can also produce some pleasant, somewhat musical, notes. The plumage of different species varies, but the dominant colours are grey, black and white, with reddish brown or chestnut maroon or bright rufous in parts for sheer effect.

The two species commonly sighted in gardens are:

Rufous-backed Shrike (*Lanius schach*), *above*
Bay-backed Shrike (*Lanius vittatus*), *below*

Shrikes

Babblers

This is a rather large group of look-alike birds. Sixteen species are found in India and five of these are certainly reported in Delhi. Of these, the following four species are commonly sighted:

> Common Babbler (*Turdoides caudatus), above left*
> Striated Babbler *(Turdoides earlei), above right*
> Large Grey Babbler *(Turdoides malcolmi), below right*
> Jungle Babbler *(Turdoides striatus), below left*

By and large, these are similar looking birds, with slight variations in size, plumage and habits. Mostly earthy brown in varying shades, with glowering eyes and loose dangly tails, all Babblers appear rather untidy and clownish. They flutter, glide and hop around in noisy small parties, sometimes called the 'seven sisters', that are chattering and chuckling all the time, while also exploring the ground and undergrowth for insects, which form the main diet. Also relish berries, fruits and flower nectar.

Nesting season is during the summer months. Cup-shaped nests are usually installed inside bushes and hedges and sometimes on trees. Their nests are regularly parasitised by the Cuckoos.

Babblers

Tawny Eagle–*Aquila rapax*
Common Hindi name: Okab
Size: 65-70 cm, slightly bigger than Black Kite
Call: Not very vocal. Makes raucous cackles like *kra* or *ke-ke-ke*

Found in open country, singly or in pairs. Essentially a scavenger feeding on carcasses, but also adept at robbing other smaller raptors of their prey. Often marauds poultry chicks.

Heavily built and has variable plumage ranging from dark brown to dirty buff, with long wings extending almost to the end of the rounded tail. Distinctive features are the flat eagle's head, strong hooked bill, powerful splayed talons and fully feathered legs. Female is slightly bigger than the male.

Nesting season is from November to April. The nest is made high on a tree. Both parents participate in raising the chicks, though the female alone incubates.

Tawny Eagle – Aquila rapax

Grey Francolin–*Francolinus pondicerianus*
Common Hindi name: Teetar
Size: 30-34 cm, smaller than domestic hen
Call: High-pitched ringing notes like *kateetar...kateetar* and *pateela...pateela*, mostly during morning and evening hours

Most common Francolin in the country. Plump and stub-tailed, with greyish brown plumage that is finely cross-barred above and below. Upper parts have long narrow pale streaks and throat has a broken black line.

Found in pairs or coveys. Largely terrestrial, though roosts on trees. Moves with agility on ground and is fast runner. Rises with whirring of wings, flying swiftly over short distances. A favourite game bird; its meat is tasty.

Prefers thorn-scrub country, but is commonly found in cultivated areas and big gardens or parks having shrubs and bushes. Feeds mainly on grain, seeds, termites and other small insects. Often scratches the ground and even cattle dung for this purpose. Lays eggs on the ground in a grass-lined scrape under a bush in scrub country. Has no specific nesting season. Breeds quite rapidly under normal circumstances.

Grey Francolin – *Francolinus pondicerianus*

Common Indian Nightjar–*Caprimulgus asiaticus*
Common Hindi name: Chapka
Size: 24-25 cm, size of Myna
Call: Typical *chuk, chuk, chuk, chuk-r-r-r,* somewhat like a stone gliding over hard ice

Nocturnal bird with soft camouflaging plumage that is mottled grey-brown and buff, with black streaks above. This pattern is well suited to the bird's habit of squatting on the ground during daytime under a bush or other cover. Becomes really active after sunset and then, through the night, its typical call can be heard at intervals. Meanwhile, it is busy hawking insects, which form the main diet.

Does not believe in making any proper nest and the eggs are laid on bare ground having some cover. Predation and trampling occurs quite often.

Common Indian Nightjar – *Caprimulgus asiaticus*

Indian Chat (also called Brown Rock Chat)–
Cercemola fusca
Common Hindi name: Dauma
Size: 17-18 cm, slightly bigger than House Sparrow
Call: Usually notes like *chee-chee, check-check*, but not very vocal

Dainty looking small bird, with plain brown plumage. Wings and tail are darker and the underside is more rufous brown. The tail is often moved up and down.

Prefers rocky areas and abandoned sites like old forts and monuments. Also seen in less disturbed parks and gardens. Feeds mostly on seeds, berries and small insects.

Nesting season is between April and August. The nest is generally made in a hole in a secluded wall or some other similar spot. Male and female share the domestic duties.

Indian Chat – Cercemola fusca

Common Iora–*Aegithina tiphia*
Common Hindi name: Shaubeegi
Size: 14-15 cm, same as House Sparrow
Call: Variety of whistling notes such as *peeou-peeou* and *wheee-choo, wheee-choo*

An attractive bird, largely greenish yellow with blackish wings and tail. Has two white bars on each wing. Both sexes look alike.

Arboreal by nature and can move swiftly among trees in gardens and parks or in forest. Feeds mostly on small insects, larvae and berries.

Generally found in pairs. Male is vocal during courtship and performs a typical display of springing into the air with plumage fluffed out to attract the female. Nesting period is from May to September. Cup-like nest of grasses is made neatly and placed in the fork of twig or branch. Both sexes share the parental duties.

Common Iora – *Aegithina tiphia*

Grey Tit (also called Great Tit)–*Parus major*
Common Hindi name: Ramgangra
Size: 13-15 cm, about the same as House Sparrow
Call: Chattering notes such as *whee-chi-chi, whee-chi-chi*

Small, active and acrobatic bird, quite attractive looking. Has smooth and glossy black head, glistening white cheek-patches, grey back with black streaks on wings. White underside has a broad black band down the centre. Bill is short and conical. Both sexes look alike.

Prefers wooded areas and generally found in pairs. Mostly insectivorous, but also takes seeds, berries and nuts. Utters chattering notes while moving about.

A cavity-nester, its nesting season extends from February to November. Sometimes two successive broods may be raised from the same nest. Male and female share the parental duties.

Grey Tit – *Parus major*

Common Chiffchaff–*Phylloscopus collybitta*
Common Hindi name: Chipchip
Size: 11-12 cm, smaller than House Sparrow
Call: Twittering notes like *chiff-chaff, chiff-chaff, peep-peep* and *wee-choo-choo*

Belongs to a group of birds called 'Leaf', because of the habit of lurking in the canopy and feeding on small insects, spiders, etc. among the foliage of trees and bushes. Essentially a winter visitor, as it breeds in the Himalayan region. Arrives in September-October; stays on till about April.

Tiny and mostly dusky or sooty in colour, with a greenish (olive green) tinge above while the underparts are dirty white. Legs and bill are black. Wings have no bars, but the restless flicking of wings is quite typical of the bird.

Common Chiffchaff – *Phylloscopus collybitta*

Common Rosefinch–*Carpodacus erythrinus*
Common Hindi name: Lal Tuti
Size: 15-16 cm, slightly bigger than House Sparrow
Call: Musical whistling notes like *tooee, tooee* and *twee-ee, tweeti-tweeti*

Breeds in the Himalayan region and comes to the plains, literally in flocks, around September, after which it stays till April. Prefers wooded country and feeds on seeds, berries, figs and also nectar and buds of certain flowers.

Male is fluffy and lovely looking. Its head, chin and throat are usually deep pink (strawberry colour), while upper parts are beige brown tinged with pink and belly is off-white but streaked pinkish. Female is less colourful—olive brown with fine streaks on head, breast and throat and two white bars on the wings. Both sexes have stout conical bill and slightly forked tail.

Due to small size and attractive plumage, the bird is coveted by the pet trade and its trapping has to be checked and prevented.

Common Rosefinch – Carpodacus erythrinus

CHECKLIST

S. No.	Family/Common Name	Scientific Name	
	Accipitridae		
1	Black Kite (*earlier* Pariah Kite)	*Milvus migrans*	65
2	Black-shouldered Kite	*Elanus caeruleus*	65
3	Brahminy Kite	*Haliastur indus*	65
4	Tawny Eagle	*Aquila rapax*	87
5	Short-toed Eagle	*Circaetus gallicus*	
6	Shikra	*Accipiter badius*	49
7	Common Kestrel	*Falco tinnunculus*	
	Alcedinidae/Halcyonidae/Cerylidae		
8	Common Kingfisher (*earlier* Small Blue Kingfisher)	*Alcedo atthis*	19
9	White-throated Kingfisher (*earlier* White-breasted Kingfisher)	*Halcyon smyrnensis*	17
10	Pied Kingfisher	*Ceryle rudis*	17

	Apodidae		
11	House Swift	*Apus affinis*	13
12	Asian Palm-Swift (*earlier* Swift)	*Cypsiurus balasiensis*	
	Ardeidae		
13	Intermediate Egret (*earlier* Median Egret)	*Mesophoyx intermedia*	67
14	Indian Pond Heron (*earlier* Paddy bird)	*Ardeola grayii*	
	Bucerotidae		
15	Indian Grey Hornbill	*Ocyceros birostris*	51
	Charadriidae		
16	Red-wattled Lapwing	*Vanellus indicus*	43
	Caparimulgidae		
17	Common Indian Nightjar	*Caprimulgus asiaticus*	91
	Columbidae		
18	Yellow-legged Green Pigeon	*Treron phoenicoptera*	45
19	Rock Pigeon (*earlier* Blue Rock Pigeon)	*Columba livia*	69
20	Oriental Turtle Dove	*Streptopelia orientalis*	

21	Little Brown Dove	*Streptopelia senegalensis*	71
22	Spotted Dove	*Streptopelia chinensis*	71
23	Red-collared Dove	*Streptopelia tranquebarica*	
	Coraciidae		
24	Indian Roller	*Coracias benghalensis*	73
	Corvidae		
25	Rufous Treepie (*earlier* Indian Treepie)	*Dendrocitta vagabunda*	47
26	House Crow	*Corvus splendens*	75
27	Jungle Crow	*Corvus macrorhynchos*	75
28	Black Drongo	*Dicrurus macrocercus*	63
29	Ashy Drongo	*Dicrurus leucophaeus*	63
30	White-bellied Drongo	*Dicrurus caerulescens*	63
31	Small Minivet	*Pericrocotus cinnamomeus*	77
32	Long-tailed Minivet	*Pericrocotus ethologus*	77
33	Scarlet Minivet	*Pericrocotus flammeus*	77
34	White-bellied Minivet	*Pericrocotus erythropygius*	77

35	Golden Oriole	*Oriolus oriolus kundoo*	31
	Cuculidae		
36	Indian Cuckoo	*Cuculus micropterus*	57
37	Asian Koel (*earlier* Common Koel)	*Eudynamys scolopacea*	59
38	Greater Coucal	*Centropus sinensis*	79
39	Common Hawk Cuckoo (*earlier* Brainfever Bird)	*Hierococcyx varius*	61
	Fringillidae		
40	Common Rosefinch	*Carpodacus erythrinus*	101
	Hirundinidae		
41	Northern House Martin	*Delichon urbica*	
42	Asian House Martin	*Delichon dasypus*	
43	Common Swallow or Barn Swallow	*Hirundo rustica*	
44	Wire-tailed Swallow	*Hirundo smithii*	
45	Red-rumped Swallow	*Hirundo daurica*	
	Irenidae		
46	Common Iora	*Aegithina tiphia*	95

106

	Laniidae		
47	Bay-backed Shrike	*Lanius vittatus*	53
48	Rufous-backed Shrike	*Lanius schach*	83
49	Common Woodshrike	*Tephrodornis pondicerianus*	
	Megalaimidae		
50	Coppersmith Barbet (*earlier* Crimson-breasted Barbet)	*Megalaima haemacephala*	37
51	Brown-headed Barbet (*earlier* Large Green Barbet)	*Megalaima zeylanica*	37
	Meropidae		
52	Little Green Bee-eater (*earlier* Small Bee-eater)	*Merops orientalis*	1
53	Blue-cheeked Bee-eater	*Merops persicus*	1
	Muscicapidae		
54	Asian Paradise Flycatcher	*Terpsiphone paradisi*	11
55	Red-throated Flycatcher	*Ficedula parva*	
56	White-throated Fantail Flycatcher	*Rhipidura albicollis*	
57	Oriental Magpie Robin	*Copsychus saularis*	9
58	Indian Robin	*Saxicoloides fulicata*	29

59	Indian Chat	*Cercemola fusca*	93
60	Common Stonechat	*Saxicola torquata*	39
61	Pied Bushchat	*Saxicola caprata*	39
62	Black Redstart	*Phoenicurus ochruros*	
63	Blue Rock Thrush	*Monticola solitarus*	7
	Motacillidae		
64	Indian Tree Pipit	*Anthus trivialis*	
	Nectariniidae		
65	Purple Sunbird	*Nectarinia asiatica*	5
	Paridae		
66	Grey Tit	*Parus major*	17
	Passeridae		
67	House Sparrow	*Passer domesticus*	27
68	Chestnut-shouldered Petronia	*Petronia xanthocollis*	27
69	Baya Weaver	*Ploceus philippinus*	33
70	Black-headed Munia	*Lonchura malacca*	

71	White-throated Munia (*earlier* Silverbill Munia)	*Lonchura malabarica*
72	Spotted Munia	*Lonchura punctulata*
73	Yellow Wagtail	*Motacilla flava* 35
74	Grey Wagtail	*Motacilla cinerea* 35
75	Citrine Wagtail	*Motacilla citriola* 35
76	White Wagtail	*Motacilla alba* 35
	Phasianidae	
77	Blue Peafowl	*Pavo cristatus* 31
78	Grey Francolin	*Francolinus pondicerianus* 38
79	Common Quail	*Coturnix coturnix*
	Picidae	
80	Yellow-crowned Woodpecker (*earlier* Yellow-fronted Pied Woodpecker)	*Dendrocopos mahrattensis* 25
81	Black-rumped Flameback (*earlier* Lesser Golden-backed Woodpecker)	*Dinopium benghalensis* 22
	Psittacidae	
82	Rose-ringed Parakeet	*Psittacula krameri* 53

109

83	Alexandrine Parakeet	*Psittacula eupatria*	53
84	Plum-headed Parakeet	*Psittacula cyanocephala*	53
	Pycnonotidae		
85	Red-whiskered Bulbul	*Pycnonotus jocosus*	15
86	Red-vented Bulbul	*Pycnonotus cafer*	15
87	White-cheeked Bulbul	*Pycnonotus leucotis*	15
	Silvidae		
88	Common Tailorbird	*Orthotomus sutorius*	23
89	Plain Prinia	*Prinia inornata*	
90	Ashy Prinia	*Prinia socialis*	
91	Common Chiffchaff	*Phylloscopus collybita*	99
92	Plain Leaf Warbler	*Phylloscopus neglectus*	
93	Common Babbler	*Turdoides caudatus*	85
94	Striated Babbler	*Turdoides earlei*	85
95	Large Grey Babbler	*Turdoides malcolmi*	85
96	Jungle Babbler	*Turdoides striatus*	85
97	Common Lesser Whitethroat	*Sylvia curruca*	

Strigidae

98	Spotted Owlet	*Athene brama*	2(
99	Barn Owl	*Tyto alba*	

Sturnidae

100	Common Myna	*Acridotheres tristis*	4(
101	Bank Myna	*Acridotheres ginginianus*	4(
102	Common Starling	*Sturnus vulgaris*	4)
103	Brahminy Starling	*Sturnus pagodarum*	4(
104	Asian Pied Starling	*Sturnus contra*	4)
105	Rosy Starling	*Sturnus roseus*	

Upupidae

106	Common Hoopoe (*earlier* Hoopoe)	*Upupa epops*	55

Zosteropidae

107	Oriental White-eye	*Zosterops palpebrosus*	3

Nomenclature follows the *Annotated Checklist of Birds of the Oriental Region* by Inskipp, Lindsey and Duckworth (1996)

Bird Family name	Species name	Species described in book

APPENDIX

Introduction

What is a bird?
A bird has been described as a 'Feathered Biped'. This description is apt and precise, and can apply to no other animal.

Birds are vertebrate, warm-blooded animals, i.e. whose temperature remains more or less constant and independent of the surrounding temperature. This is in contradistinction to Reptiles, Amphibians and Fishes which are cold-blooded, i.e. of temperature that changes with the hotness or coldness of their surroundings.

To assist in maintaining an even temperature, the body of a bird is covered with non-conducting feathers—its chief characteristic—which in details of structure and arrangement reflect the mode of life of the group to which the bird belongs. Compare for example the thick, soft, well-greased covering on the underside of an aquatic bird like a Duck or Grebe with the particular, narrow, hairlike, 'double' feathers of the Cassowary to be seen in any zoo. Except in the flightless birds such as the last named, the Ostrich and the Penguin *(Ratitae* and *Sphenici)* whose feathers grow more or less evenly over the entire surface of the body, the growth of feathers is restricted only to well-defined patches or tracts known as *pterylae* on various parts of the body, whence they fall over and evenly cover the adjoining

---

From Salim Ali, *The Book of Indian Birds*, 13th ed., 2002, pp. xviii-xxv (Introduction) and pp 306-14 ('The Usefulness of Birds' and 'Birdwatching'), first published in 1941.

naked interspaces or *apteria*. A study of the arrangement of the feather tracts (*pterylosis*) which varies in the different orders, families, and even species, is of great importance in determining the natural relationships of different birds.

The feathers covering the body of a bird fall into three classes: (1) the ordinary outside feathers known as contour feathers or *pennae*, whether covering the body as a whole or specialised as pinions or flight feathers (*remiges*) or as tail feathers (*rectrices*) which serve as rubber and brake; (2) the fluffy Down feathers or *plumulae* hidden by the Contour feathers and comparable to flannel underclothing, whether confined to nestlings or persisting throughout life; (3) the hair-like Filo-plumes which are hardly seen until the other feathers have been plucked off. They are particularly noticeable, for instance, in a plucked pigeon.

The body temperature of birds, about 38°-44°C, is higher than that of most mammals. Assisted by their non-conducting covering of feathers, birds are able to withstand great extremes of climate. As long as they can procure a sufficiency of food supply, or 'fuel' for the system, it makes little material difference to them whether the surrounding temperature is over 60°C, on the burning desert sands or 40°C, below zero in the icy frozen north. Their rate of metabolism is higher than that of mammals. They lack sweat glands. The extra heat generated by their extreme activity which would, under torrid climatic conditions result in overheating, fever, and death, is eliminated through the lungs and air sacs as fast as it is produced. For one of the functions of the 'air sacs'—a feature peculiar to birds and found in various parts within the body—is to promote internal perspiration. Water vapour diffuses from the blood into these cavities and passes out by way of the lungs, with which they are indirectly connected.

In addition to these two cardinal attributes, warm-bloodedness and insulated feather covering, birds as a class possess certain well-marked characteristics which equip them pre-eminently for a life in the air. In India we have at present no indigenous flightless birds like the Ostrich or the Penguin, so these need not be considered here. The forelimbs of birds, which correspond to human arms or to the forelegs of quadrupeds, have been evolved to serve as perfect organs of propulsion through the air. Many of their larger bones are hollow and often have air sacs running into them which, as mentioned above, function principally as accessory respiratory organs. This makes for lightness without sacrificing strength, and is a special adaptation to facilitate aerial locomotion. Modification in the structure of the breastbone, pectoral girdle and other parts of the skeleton, and the enormously developed breast muscles enable a bird to fly in the air. It has been estimated from analogy with birds that a man, to be able to lift himself off the ground by his own effort, would require breast muscles at least 1.2m deep! There is, moreover, a general tendency for various bones to fuse with each other, conducing to increased rigidity of the skeletal frame—also a factor of great importance in flight. As a whole the perfectly streamlined spindle-shaped body of a bird is designed to offer the minimum resistance to the wind. On account of their warm-bloodedness coupled with these peculiar facilities for locomotion with which Nature has endowed them, birds enjoy a wider distribution on the earth than any other class of animals. They cross ocean barriers and find their way to remote regions and isolated islands, and exist under physical conditions where their cold-blooded relatives must perish. It is also this power of swift and sustained flight that enables birds living in northern lands to migrate periodically over enormous distances in order to escape from the rigours of winter—shortening days and dwindling food supply—to warmer and more hospitable climes.

Birds are believed to have sprung from reptilian ancestors in bygone aeons. At first sight this appears a far-fetched notion, for on the face of it there seems little in common between the grovelling cold-blooded reptile and the graceful, soaring, warm-blooded bird. But palaeontological evidence, supplied chiefly by the earliest fossil of an undoubted bird to which we have access—the Archaeopteryx—and modern researches on the skeletal and other characteristics of our present-day birds, tend in a great measure to support this belief. The method of articulation of the skull with the backbone, for instance, and the nucleated red blood corpuscles of the bird are distinctly reptilian in character. To this may be added the fact that birds lay eggs which in many cases closely resemble those of reptiles in appearance and composition, and that the development of the respective embryos up to a point is identical. In the majority of birds scales are present on the tarsus and toes which are identical with the scales of reptiles. In some birds, like Sandgrouse and certain Eagles and Owls, the legs are covered with feathers, a fact which suggests that feathers are modified scales and that the two may be interchangeable. The outer covering of the bills of certain birds, for example the Puffin (*Fractercula arctuca*), is shed annually after breeding in the same way as the slough in reptiles. The periodical moulting of birds is also essentially the same process as the sloughing of reptiles. In short, birds may reasonably be considered to be extremely modified reptiles, and according to the widely accepted classification of the great scientist T.H. Huxley, the two classes together form the division of vertebrates termed Sauropsida.

* * * * *

Of the senses, those of sight and hearing are most highly developed in birds, that of taste is comparatively poor, while smell is practically absent. In rapid accommodation of the eye, the bird surpasses all other creatures. The focus can be altered from a distant object to a near one almost instantaneously; as an American naturalist puts it, 'in a fraction of time it (the eye) can change itself from a telescope to a microscope'.

* * * * *

For the safety of their eggs and young, birds build nests which may range from a simple scrape in the ground, as of the Lapwing, to such elaborate structures as the compactly woven nest of the Weaver Bird. With rare exceptions they incubate the eggs with the heat of their own bodies and show considerable solicitude for the young until they are able to fend for themselves. Careful experiments suggest however, that in all the seemingly intelligent and purposeful actions of nesting birds, in the solicitude they display for the welfare of their young and in the tactics they employ when the latter are in danger, *instinct* and not intelligence is the primary operating factor. The power of reasoning and the ability to meet new situations and overcome obstacles beyond the simplest, are non-existent. It is good therefore always to bear this in mind when studying birds, and to remember that their actions and behaviour cannot be judged purely by comparison with human standards and emotions.

*   *   *   *   *

A few remarks with regards to the classification of birds seem called for in the interest of the beginner. It will be observed that after the English or trivial name of each species in the following pages, there appear two Latin names. The practice of employing a uniform Latin terminology is current throughout the modern scientific world. It is a boon to workers in different countries since it is more or less constant and enables the reader of one nation to understand what the writer of another is talking about. To take an example: What the Englishman calls Hoopoe is Wiedehopf to the German. A Pole knows the bird as something else—doubtless with a good many c's, z's, s's and other consonants in bewildering juxtaposition—while the Russian has yet another equally fantastic looking name for it. A fair working knowledge of a language seldom implies a familiarity with popular names as of birds, for instance, many of which often are of purely local or colloquial applications. Thus it is possible that while the Englishman may follow more or less all he reads in German about the Wiedehopf he may still be left in some doubt as to the exact identity of the bird. The international Latin name *Upupa epops* after the English or Polish or Russian name will leave no doubt as to what species is meant.

In the above combination the first name *Upupa* denotes

the Genus of the bird corresponding roughly, in everyday human terms, to the surname. The second name *epops* indicates the Species and corresponds, so to say, to the Christian name. Modern trend of scientific usage has tended to split up the Species further into smaller units called Geographical Races or Subspecies. An example will clarify what this means: It will be admitted that all the peoples living in India are human and belong to one and the same human species. Yet a casual glance is enough to show that the Punjabi is a very different type in build and physiognomy from the dweller in Madras. The differences, though small, are too obvious to be overlooked. They are primarily the result of environment which includes not only climate conditions of heat and cold, dampness, but also of diet and many other subtle factors working unceasingly upon the organism in direct or indirect ways. Thus while retaining all our inhabitants under the human species, when you talk of the Madrasi or the Punjabi you automatically recognize the sum total of the differences wrought in either by his particular environment.

A comparative study of birds reveals that there are similar minor but well-marked and readily recognizable differences in size, colouration and other details in those species which range over a wide area and live under diversified natural conditions, or which have been subjected to prolonged isolation as on oceanic islands, or through other causes. It is important for science that those differences should be duly recognized and catalogued since they facilitate the study of speciation and evolution. This recognition is signified by adding a third Latin name to the two already existing, to designate the Geographical Race or Subspecies. Thus, for example the House Crow, *Corvus splendens*, has been subdivided on the basis of constant differences in size and colouration brought about in the different portions of the 'Indian Empire' it occupies as follows:

Corvus splendens splendens (the nominate race), Common House Crow
Corvus splendens zugmayeri, Sind House Crow

Corvus splendens insolens, Burmese House Crow
Couvus splendens protegatus, Ceylon House Crow

Barring restricted areas and particular groups of birds which still require careful collecting and working out, we can now claim to have a sufficiency of dead ornithological material from India in the great museums of the world to satisfy the needs of even an exacting taxonomist. Most bird lovers in this country possess neither the inclination, training, nor facilities for making any substantial additions to our knowledge of systematics. Speaking generally, therefore, Indian *systematic* ornithology is best left in the hands of the specialist or museum worker who has the necessary material and facilities at his command. Our greatest need today is for careful and rational field work on *living* birds in their natural environment, or what is known as Bird Ecology. It is a virgin field; both the serious student and the intelligent amateur can contribute towards building up this knowledge. A great many biological problems await solution by intensive ecological study. This is a line of field research that may be commended to workers in India; it will afford infinitely more pleasure and is capable of achieving results of much greater value and usefulness than the mere collecting and labelling of skins.

Among the questions which the ornithologist in India is constantly being asked are the following. I have had to face them so frequently, from such a variety of people and in such far-flung corners of the country that it might perhaps be as well to devote a little space to them here.

Q. *What is the largest Indian bird, and what the smallest?*

A. It is not easy to say which particular one is the largest, but amongst the upper ten are certainly the Sarus Crane and the Himalayan Bearded Vulture or Lammergeier. The former stands the height of a man; the latter has a wingspread of over 8 feet. Amongst our smallest birds are the flowerpeckers, e.g.

Tickell's Flowerpecker is scarcely bigger than a normal thumb.

Q. What is our most beautiful bird?

A. Difficult to pick out any single species for the highest honour, and depends rather on individual tastes. A large number of birds of many different families, particularly those resident in areas of humid evergreen forests, possess extraordinarily brilliant plumage. As a family, the Pheasants occupy a high place for colour and brilliancy of plumage and adornment possessed by the cocks of most species. At the bottom of the size ladder come the Sunbirds—tiny creatures about half the size of the House Sparrow or less—whose glistening resplendent plumage scintillating in the bright sunshine as they flit from flower to flower, or dart from one forest glade to another, transforms them into living gems.

Q. What is our commonest bird, and what is our rarest?

A. The answer depends largely on what part of the country you live in. But for India as a whole, perhaps the House Crow and the House Sparrow would be hard to beat for commonness and abundance. They have followed man everywhere—up in the hills and out in the desert—wherever his ingenuity has created liveable conditions for himself. Next in abundance come birds like Mynas and Bulbuls which though not wholly commensal on man are yet quick to profit by his presence and activities.

Perhaps the three rarest birds in India at present are the Himalayan Quail *(Ophrysia superciliosa)*, Jerdon's Courser *(Rhinoptilus bitorquatus)*, and the Pinkheaded Duck *(Rhodonessa caryophyllacea)*... The first has not been met with since 1876 and all attempts to re-discover them have ended in failure. The second which had not been seen since 1900 was rediscovered in the Lankamalai Sanctuary of Andhra Pradesh in 1986 by Dr Bharat Bhushan, a young scientist of the BNHS working on a project on endangered and rare species of birds.

The fate of the Pinkheaded Duck is also shrouded in mystery and to all appearances the species has become extinct. The last example shot was in 1935, and though it has since been reported off and on as seen by sportsmen, in all the cases investigated these reports have proved unreliable, the bird actually seen being the Redcrested Pochard.

Q. Do birds have a language?

A. They certainly have, if by language is meant that they can communicate with and understand one another. It consists not of speech as we know it, but of simple sounds and actions and enable birds—specially the more sociable ones—to maintain contact amongst themselves and convey simple reactions such as those of pleasure, threat, alarm, invitation, and others. Several of these signals—vocal, behavioural, or a combination of the two—are understood not only by members of the same species but also by other birds generally, e.g. the alarm-notes and behaviour of many on the approach of a marauding Hawk. To this extent man can also claim to understand the language of birds; Solomon himself could hardly have done more. But the structure of a bird's brain suggests a comparatively low level of intelligence and precludes the possibility of their holding regular conversations or expressing views and opinions as we humans are usually so ready to do!

Q. What is our most accomplished songster and talker?

A. Personally, for song I would give the palm to the Greywinged Blackbird (*Turdus boulboul*) of the Himalayas. A number of its close relations, members of the Thrush family, including the Malabar Whistling Thrush and the White-rumped Shama follow close on its heels.

The best talker amongst our Indian birds is certainly the Common Hill Myna whose articulation of the human voice and speech is infinitely clearer and truer than that of the Parakeets.

The latter enjoy a wider reputation and are more generally kept as cage birds because they can be more readily procured.

Q. How long does a bird live?

A. The age-potential, or the age to which a bird is capable of living, of course varies according to the species and the environment and conditions under which it lives. Reliable data concerning the life-span of wild birds in a state of nature are very difficult to obtain. It is only possible by the method of marking individual birds, particularly as nestlings. Most of the figures of age available are from birds in captivity and therefore living under somewhat unnatural conditions. It is known that within a group of related forms the larger the birds the longer its life, but outside related groups size does not seem to matter a great deal. An ostrich in captivity has lived for 40 years; a Raven to 69 and another to 50. Passerine birds of about Sparrow size have occasionally reached 25 years, although normally their span is 5 to 8. A vulture attained 52, a horned owl 68, swan 25, pigeon 22 to 35, peacock 20. The longest lived wild birds in a natural state, as determined by the marking method, are: herring gull 36 years, oriole 8, pintail duck at least 13, grey heron about 16, blackbird 10, curlew 31½, kite 25¾, and swallow 16+.

The common belief that Crows are immortal is of course groundless, while there seems no proof for the popular assertion that vultures 'score centuries'...

THE USEFULNESS OF BIRDS

It has been said that birds could exist without man but that man would perish without birds. This observation has been further amplified by the remark that 'but for the trees the insects would perish, but for the insects the birds would perish, but for the birds the trees would perish, and to follow the inexorable laws of Nature to the conclusion of their awful vengeance, but for the trees the world would perish'. An impartial scrutiny of the facts shows that there is indeed little extravagance in either of these speculations.

As destroyers of insect pests

The variety, fecundity and voracity of insects are unbelievable. Over 30,000 forms have been described from the Indian region alone—about fifteen times the number of bird species and races—and probably many still remain to be added to the list. Practically all living animals as well as plants furnish food to these incomputable hordes. Many estimates have been made of what a single pair of insects would increase to if allowed unchecked multiplication, and astounding figures have been reached rivalling in their stupendousness those which we associate with astronominal calculations. A Canadian entomologist has estimated that a single pair of Colorado Beetles or Potato Bugs (*Leptinotarsa decemlineata*—belonging to the prolific family Chrysomelidae of which over 20,000 species are known throughout the world, and which is well represented in India) would, without check, increase in one season to sixty million. Riley computed that the Hop Aphis or Chinch Bug (*Blissus leucopterus*), very destructive to grasses and cereals in America, which develops 13 generations in a single year would, if unchecked, reach ten sextillion individuals at the end of the 12th generation. If this brood were marshalled in line end to end at the rate of 10 per inch, the procession would be so long that light, travelling at the rate of 2,95,000 km per second, would take 2,500 years to reach from one end to the other!

A caterpillar is said to eat twice its own weight in leaves per day. Certain flesh-feeding larvae will consume within 24 hours 200 times their original weight. It is reckoned that the food taken by a single silkworm in 56 days equals 86,000 times its original weight at hatching. Locusts are as notorious for their prolific reproduction as for their prodigious appetites. Their swarms are sometimes so thick as to obscure the sun, and such a visitation will, in the course of a few short hours, convert a green and smiling tract into a desolate waste with nothing but bare stems. The female locust lays its eggs in capsules underground, each capsule containing about 100 eggs, and several of these capsules are laid by each individual. On a farm in South Africa measuring 1335 ha, no less than 14 tons of eggs have been dug up at one time, estimated to have produced 1,250 million locusts. It is evident from their rate of increase that unless insect numbers were kept under constant and rigid check, it would not be long before all vegetation vanished completely from the face of the earth.

A large proportion of the normal food of birds consists of insects, including many that are in the highest degree injurious to man and his concerns. Birds of many species not only take heavy toll of the marauding locust hordes all along their flight lines, but also scratch up and devour their eggs in vast quantities, as well as the different stages of the young locust after hatching. The White Stork is a well-known locust destroyer, and the enormous nesting colonies of the Rosy Starling live and feed their young exclusively upon these insects on their common breeding grounds in Central Asia. An idea of the extent of good birds do in destroying insect pests may be had from the fact that many young birds in the first few days of their lives consume more than their own weight of food in 24 hours. A pair of Starlings have been observed to bring food (caterpillars, grasshopper, locusts, etc.) to their nest/young ones 370 times in a day, and according to Dr W.E. Collinge, the well-known British authority, House Sparrows bring food (caterpillars, soft-bodied insects, etc.) from 220 to 260 times per day. A German ornithologist has estimated that a single pair of Tits with their progeny destroys annually at least 120 million insect eggs or 1,50,000 caterpillars and pupae. This warfare is waged not only when the insects are at the peak of their periodical abundance, but incessantly, relentlessly, and in all stages of the insects' lives. Therefore, where birds have not been unwisely interfered with, they constitute one of the most effective natural checks upon insect numbers.

As destroyers of other vermin

Owls, Kestrels, Hawks and the birds of prey generally—so often accused of destructiveness to poultry and game and slaughtered out of hand—are amongst the most important of nature's checks upon rats and mice, some of the most fecund and destructive pests from which man and his works suffer. These vermin do enormous damage to crops and agricultural produce, and are, besides, the carriers, directly or indirectly, of diseases often fatal to man. The ravages of the Sind Mole-rat in the rice-growing tracts of the Indus Delta in Lower Sind have been estimated by a competent investigator as between 10 and 50 per cent of the entire paddy crop. This Mole-rat breeds throughout the year. The number of young born in a litter is 5 to 10, but in October and November the litters are very large, varying from 14 to 18 young each. Mice are equally fecund and destructive.

It has been computed that one pair of House Rats having 6 litters of 8 young annually and breeding when 3.5 months old, with equal sexes and no deaths, would increase at the end of the year to 880 rats. At this rate the unchecked increase of a pair in five years would be 940,369,969,152 rats. Such calculations, of course, are purely theoretical and their results will never be approached in nature, but they are not extravagant considering the capacity to reproduce, and are based on moderate and conservative estimates.

It will thus be seen that every pair of rats destroyed by birds means the annual suppression of a potential increase of 880 rats. Many of our Owls and diurnal birds of prey feed largely on rats and mice; some of the former, indeed, live more or less exclusively on them. Two or three rats or mice apiece, or their remains, may frequently be found in the stomachs of Eagle-owls, for example, and as digestion in birds is a continuous and rapid process it is conceivable that a larger number may be taken in the course of 24 hours. Since these birds are engaged in the good work from the year's end to year's end, some estimates of their beneficial activities can be made.

As scavengers

Vultures, Kites and Crows are invaluable scavengers. They speedily and effectively dispose of carcasses of cattle and other refuse dumped in the precincts of our villages—notoriously lacking in any organised system of sanitation—that would otherwise putrefy and befoul the air and become veritable culture beds of disease. The services of the birds are of special importance during famines and cattle epidemics when large numbers of domestic animals perish and at best are left by the wayside covered with a flimsy layer of earth to be exhumed by the first prowling jackal that happens on the spot. The speed and thoroughness with which a party of Vultures will dispose of carrion is astounding.

As flower-pollination agents

While the importance of bees, butterflies and other insects in the cross-fertilisation of flowers is well known, the large part played by birds in the same capacity has not been adequately appreciated. A large number of birds of diverse families and species are responsible for the cross-fertilisation of flowers, many of them possessing special adaptations in the structure and mechanism of their tongue and bill for the purpose of extracting honey from the base of the flower tubes. Flower-nectar is rich in carbohydrates and provides excellent nutriment, so much so that many of the most highly organised flower-birds subsist more or less exclusively on this diet. In trying to reach the nectar, the forehead or throat of the bird comes into contact with the anthers. The ripe golden pollen dust adheres to the feathers and is transported to the mature stigma of the next flower visited, which it thus fertilises. It is little realised how largely responsible birds are for the success of the present-day safety match industry in India. Of all the indigenous softwoods that have been tried in the manufacture of matches, that of the silk cotton tree has been found to be the most satisfactory as regards quality, abundance and accessibility. The large showy crimson flowers of this tree serve as a sign-post to invite the attention of the passing bird. They contain a plentiful supply of sugary nectar, which is eagerly sought by birds of many kinds—over 60 different species have been noted in one small area alone— and are mainly cross-pollinated through their agency. Birds thus contribute to the production of fertile seed and the continuance of healthy generations of the tree, and incidentally to the supply of raw material for your box of matches. A careful scrutiny would reveal that we are ultimately dependent upon birds in this house-that-Jack-built sort of way for many more of our everyday requirements. The Coral tree (*Erythrina*), which is largely grown for shade in the tea and coffee plantations of South India, is also one whose flowers are fertilised chiefly, if not exclusively, by birds of many species.

As seed dispersers

In the dissemination of seed and the distribution of plant life, birds play a predominant part. Their activities unfortunately, are not always of beneficial character from the economic point of view. No better instance of the extent of their seed-dispersing activities can be cited than that of the lantana weed. This pernicious plant of Mexican origin was first introduced into Sri Lanka for ornamental purposes a little over a century ago. It has since overrun thousands of square miles of the Indian subcontinent, and become the despair of agriculturist and forester alike. Its phenomenal spread within this comparatively short period would have been impossible without the agency of birds, numerous species of which greedily devour the berries which the plant everywhere produces in such overwhelming profusion. A Black-headed Oriole has been observed swallowing 77 berries in the course of 3 minutes. The seeds pass through the birds' intestines unaffected by the digestive juices, and out with the waste matter in due course. They germinate rapidly under favourable conditions and establish themselves.

Another noxious plant that is entirely bird-propagated is the *Loranthus* (=*Dendrophthoe*) tree-parasite. It belongs to the Mistletoe family, well represented in this country, almost all of whose Indian members are more or less wholly symbiotic with Sunbirds, Flowerpeckers and other bird species, which both fertilise its flowers and disperse its seeds. Bulbuls and Barbets are largely responsible for the dissemination of the seeds of the sandalwood tree in South India and are welcome in sandalwood plantations. In the newly colonised canal areas of Punjab, the mulberry owes its abundance mainly to propagation by birds. Experiments have shown that the seeds of such plants which grow on richly manured soil, after passing uninjured through a bird's intestine, actually produce stronger seedlings than those which are cultivated without such treatment.

As food for man

A feature of the larger *dhands* or *jheels* in Pakistan and northern India during the cold weather is the magnitude of the netting operations that go on throughout that season for supplying the markets of the larger towns, both near and distant, with Wildfowl of every description for the table. The population of the neighbourhood of those *jheels* subsists during those months more or less exclusively on the flesh of water birds or in the traffic in them. Round every village near a *dhand* of any size in Sind may be seen little mounds of Coot feathers which furnish an indication of the esteem which the bird enjoys as an article of diet. The Wildfowl netting operations on the Manchar Lake alone involved, in pre-Partition days, a turnover of several thousand rupees annually, besides providing the inhabitants of the neighbourhood with free or almost free sustenance for several months in a year.

Quails, Partridges and other game birds are also netted or shot for eating purposes, and innumerable other species of every description are captured and sold in the bazaars of fanciers or exporters, yielding substantial returns to those engaged in the trade.

Egret feathers

Until a few years ago Egret farming for the sake of the valuable plumes was a profitable cottage industry and largely practiced on the various *jheels* in Sind. The dainty 'decomposed' breeding plumes of the White Egrets—'aigrettes' as they are known to the trade—were largely exported to Europe for ladies' head dresses, tippets, muffs and for other ornamental purposes. They were almost worth their weight in silver, and brought in handsome profit to the farmers. With the change in ladies' fashions, the demand has happily dwindled considerably, and with it the prices. The working of the Wild Birds and Animals Protection Act of 1912 imposed a further check upon exports, and most of the Egret farms have now disappeared.

Birds' nests

There are other minor products of birds which, if properly husbanded could be made to yield sizeable revenue in India. The saliva nests of the so-called Edible-nest Swiftlets (*Collocalia*), which breed in vast colonies in grottoes on the rocky islands off the south Myanmar coast were a source of considerable income to those engaged in the trade, and of royalty to government before that country was separated from India—and doubtless still are. These Swiftlets also breed on certain islets off the Konkan coast (W. India), but the nests here are of poor quality; the trade in them, which was small even in former years, is now non-existent. The nests were exported to China as an epicurean delicacy, the better qualities fetching from Rs 15 to 30 per kg. The value of nests imported into China during 1923, 1924 and 1925 exceeded Rs 25 lakhs; a fair proportion of these came from the then British Indian Empire.

Guano

Guano which is really the excrement of sea birds such as Gannets, Cormorants and Pelicans is another product of great commercial value. The fertilising properties of the phosphoric acid and nitrogen contained in fish were not recognised until guano became a stimulus to intensive agriculture. The real guano is found in vast stratified accretions on rainless islands off the coast of Peru, and although no deposits of like magnitude or value exist within our limits, yet the possibilities of the 'liquid guano' of colonial-nesting water birds have not been seriously exploited in India.

From all that has been said it must not be assumed that birds are a wholly unmixed blessing. They are injurious to man's interest in a number of ways. They destroy his crops, and damage his orchards, flower beds and vegetable gardens; they devour certain beneficial insects and prey upon fish and other animals useful to man as food; they act as intermediate hosts of parasites and viruses that spread disease among his livestock, and disperse them far and wide in the course of their migrations; they fertilise the flowers and disseminate the seeds of noxious plants and weeds. Yet, considering everything, there can be no doubt that the good they do far outweighs the harm, which must therefore be looked upon as no more than the labourer's hire.

The case for the protection and conservation of birds in a country like ours—so largely agricultural and forested and therefore at their mercy—is clear, and needs no eloquent advocacy. Quite apart from the purely materialistic aspect, however, it must not be forgotten that man cannot live by bread alone. By the gorgeousness of their plumages and the loveliness of their forms, by the vivaciousness of their movements and the sweetness of their songs, birds typify Life and Beauty. They rank beyond a doubt among those important trifles that supplement bread in the sustenance of man and make living worthwhile.

BIRD WATCHING

Nearly everyone enjoys birds: the beauty of their forms and colouring, the vivacity of their movements, the buoyancy of their flight and the sweetness of their song. It is precisely on this account that as pursuit for the out-of-doors, bird watching stands in a class by itself. Its strong point is that it can be indulged in with pleasure and profit not only by the man who studies birds scientifically, but also by one possessing no specialised knowledge. The latter, moreover, is enabled to share his profit with the scientist who for certain aspects of bird study has to depend entirely upon data collected by the intelligent watcher.

The appreciation of the beautiful and the novel is a characteristic latent in the human species. There is none in whom the seed of this faculty is entirely wanting. Environment may nurture and develop it in some, smother it in others. The fact of its existence is proved by the enquiries an ornithologist frequently receives concerning the identity of this bird with a green head or that with a red tail from persons of the most prosaic 'butcher, baker and candlestick-maker' type who in the course of their day-to-day lives would never dream of going a step out of their way merely to look at a bird. It shows that even such a person, in spite of himself, cannot at one time or another help being struck by some peculiarity in the sight or sound of a bird which had not forced itself on his notice before.

It is amazing what tricks the imagination can play with undisciplined observation. A person who, for example, notices a male Paradise Flycatcher for the first time and is struck by its exquisite tail-ribbons fluttering in the breeze, will, as likely as not, and in all good faith, clothe his bird in multi-coloured hues of green and blue and yellow and red when describing it to you. The only real clue he furnishes is the ribbon tail. Some days later you have an opportunity of pointing out a Paradise Flycatcher to your enquirer with a suitable suggestion, whereupon you promptly learn that this indeed was the object of his ecstasy!

Yet it is equally amazing what small effort is needed to discipline oneself to observe accurately. After a comparatively short period of intelligent bird watching one can often become so proficient that the mere glimpse of a bird as it flits across from one bush to another—some distinctive flash of a colour, a peculiar twitch of the tail—is enough to suggest its identity fairly reliably. If it is an unfamiliar species this fleeting impression will often suffice to puzzle it out with the aid of a bird book afterwards.

Apart from the joy and exhilaration it affords, careful and intelligent bird watching—considering that it can be indulged in by the many without special scientific training—widens the scope immensely for procuring data relating to the lives and behaviour of birds. Observations by people who habitually watch birds even merely for pleasure, are often of great value to the scientist trying to unravel some particular phase of bird life. Indeed, such observations—made as they are without knowledge of, or being swayed by this pet theory or that—frequently carry the added advantage of being completely unbiased. As mentioned in a previous chapter the bulk of the work that now remains to be done on the birds of India concerns the *living* bird in its natural surroundings: How does the bird live and behave? In what way is it fitted or is fitting itself to its habitat? How is it influenced by or is influencing its environment? It is only satisfactory answers to questions like these—and their number is legion—that can lead us to a better understanding of that very real but strangely elusive thing called life.

One often hears it asserted that there are no birds in this locality or that. Such statements merely suggest that the observer may not know exactly where and how to look for them. For indeed it is difficult to imagine a single square mile of the Indian subcontinent entirely devoid of birds. Even in the midst of the scorching Rajasthan desert or amongst the high Himalayan snows, birds there are for those who know how to find them. They must be scarce and local, simply because their food happens to be scarce and local, but they are never entirely absent over areas of any size.

For the new arrival in this country and for the novice, some suggestions as to when and where to look for birds with success might prove helpful. First and foremost, although birds are on the move all day long, their activity is greatest in the early mornings; therefore early rising is a most important pre-requisite for successful watching. Most song is also heard during the early morning hours. Discovering the identity of a songster often entails patient watching, and the chances of tracking him down are naturally greatest in the early morning when the bird is most vocal.

Contrary to the popular notion, a forest, to the inexperienced, is usually a very disappointing place for bird watching. You may tramp miles without seeming to meet or hear a bird, and then just as you begin to despair you may round a bend in the path and suddenly find yourself confronted by a gathering that includes well nigh every species of the neighbourhood. There are birds on every hand: on the ground, among the bushes, on the trunks of the lofty trees and in the canopy of leaves high overhead. There are tits, babblers and tree pies, woodpeckers, nuthatches and drongos, flycatchers, minivets, and leaf-warblers, and numerous other species besides. The scene is suddenly transformed into one of bustling activity. You have in fact struck what the books call a 'Mixed Hunting Party', or 'Localised Forest Association'. These mixed assemblages are a characteristic feature of our forests, both hill and plain. Here birds do not as a rule spread themselves out uniformly, but rove about in cooperative bands of mixed species in more or less regular daily circuits. All the members of the association profit through the co-ordinated efforts of the lot. Babblers rummaging amongst the fallen leaves for insect food disturb a moth which is presently swooped upon and captured in mid-air by a drongo on the look-out hard by. A woodpecker scuttling up a tree-trunk in search of beetle galleries stampedes numerous winged insects resting upon the protectingly coloured bark or lurking within its crevices. These are promptly set upon by a vigilant flycatcher or warbler—and so on.

Banyan and *peepul* trees when in ripe fig attract a multitude of birds of many species from far and wide and offer excellent opportunities to the bird watcher. A lively scene presents itself as a party after party arrives, all eager to gorge themselves on the abundance spread around. There is a great deal of noise and chatter as the visitors hop from branch to branch in their quest. Bickering and bullying are incessant, but no serious encounters develop since every individual is much too preoccupied with the main business in hand. Such gatherings are ideal for studying the natural dispositions and 'table manners' of the various species.

Some of the most charming and enjoyable venues for bird watching are certainly afforded by the Silk Cotton, Coral Flower, or Flame-of-the-Forest (*Butea*) trees in bloom. Their particular attractiveness lies in the fact that the trees, or the branches bearing the gorgeous flowers, are bare and leafless at this season, allowing a clear and unobstructed view of the visitors. Almost every small bird of the surrounding countryside flocks to the blossoms for the sake of the sugary nectar which they produce in such abundance. Riot and revelry prevails throughout the day, but especially in the mornings, and there is constant bullying, hustling and mock fighting amongst the roysterers. A pair of good binoculars multiplies the pleasure of bird watching manifold, and is indeed an indispensable item of the watcher's equipment.

Another favourable occasion is after the first few showers of rain have fallen and the winged termites—the potential queens and their numerous suitors—are emerging from their underground retreats for their momentous nuptial flight. A termite swarm acts like a magnet upon the bird population of its neighbourhood. Caste and creed are forgotten and every species hastens to the repast; no quarter is given, the insects being chased and captured on the ground as well as in the air. The agile and graceful gliding swoops of the Swallows and Swifts contrast strangely with the ponderous, ungainly efforts of Crows making unaccustomed aerial sallies in the pursuit. Kites, kestrels, crows, owlets, mynas and bulbuls, sparrows, bayas and munias, treepies, drongos and orioles, tailorbirds and prinias, all join in the massacre, while even woodpeckers and barbets can seldom resist the temptation...

Everyone who watches birds intelligently enough then, and who carries with him a notebook and pencil, should be in a position to contribute in some measure to our store of knowledge. The essentials are patience, plus the ability to observe accurately and to record faithfully, even though the observations may sometimes disagree with the books or the observer himself may sometimes wish things to happen differently!

Above all it is important that sentimentality be kept in check and to remember at all times that the behaviour of birds cannot be interpreted entirely by human analogy. Birds do not possess the power of reasoning: therefore their actions, however intelligent they may seem, are essentially little more than instinctive reflexes...

*   *   *   *   *